# Lighting, Audio & Studio Setup on Any Budget

From Natural Light to Full Studio, Create the Perfect Setup at Home

*By Joe Correa*

Published by Live Stronger Faster



This book is for educational and informational purposes only. The author and publisher are not engaged in rendering legal, financial, or other professional advice. Readers should consult a qualified professional before making any financial decisions.

First Edition

# Table of Contents

Introduction........................................................................5

Chapter 1: Mastering Natural Light Before Anything Else................11

Chapter 2: Building Your First Budget Lighting Kit......................... 21

Chapter 3: Audio Quality First: Choosing the Right Microphone on Any Budget........................................................................ 30

Chapter 4: Transforming Any Room Into a Recording Space..........38

Chapter 5: Camera Positioning, Angles & Framing for a Professional Look.................................................................................. 46

Chapter 6: Eliminating Echo, Noise & Reverb: How to Sound Studio-Quality Anywhere................................................................ 55

Chapter 7: Mastering Backgrounds, Depth & Visual Aesthetics...... 63

Chapter 8: Budget Tripods, Mounts & Stabilization for Smooth, Professional Shots.............................................................................. 71

Chapter 9: Color Temperature, White Balance & Matching Your Lights for Professional Consistency................................................ 79

Chapter 10: Building a Full Home Studio Step-by-Step (Beginner, Intermediate & Advanced Setups).................................................. 88

Epilogue.................................................................................. 96

# Introduction

Creating beautiful, professional-quality videos no longer requires a Hollywood studio, expensive cameras, or an elaborate production team. Today, creators all around the world are building powerful brands and earning full-time incomes from setups that start with nothing more than a sunny window, a smartphone, and a little creativity. Lighting and audio used to be complex technical fields reserved for filmmakers and sound engineers; now, they are accessible to anyone willing to learn the foundational principles. This book was created to show you that you can achieve the same cinematic results as top creators—whether you have a $0 budget or you're ready to evolve into a fully optimized home studio setup. In the world of online video, lighting and audio quality are not nice-to-haves; they are the backbone of your viewer's experience, the first impression of your professionalism, and one of the most important factors determining whether someone continues watching your content. Your audience may tolerate a shaky camera or a basic backdrop, but poor lighting and muffled audio are unforgivable. The good news is that both can be fixed faster than almost any other part of the production process. A great lighting setup can be achieved in minutes. A crisp audio setup can be added with a single plug-and-play device. And when combined, those two upgrades can instantly elevate your content from amateur to high-authority expert.

This book is designed as your complete roadmap, covering everything from natural lighting to advanced studio configurations, so you can create high-quality videos regardless of your environment, space limitations, or financial resources. You will learn how to master lighting using nothing but your window and a reflector before scaling into ring lights, softboxes, LED panels, and full three-point lighting systems. You'll understand how to analyze your room for echo, reverb, and noise so you can create the cleanest possible audio environment—even if you live in a busy home with kids, pets, or street noise. You will discover the secret strategies used by top creators to produce crisp, professional sound using simple budget microphones as well as how to upgrade strategically as your channel, brand, or income grows. Whether you film in a bedroom, kitchen, garage, office, or even your car, you'll find step-by-step guides to transform that space into a production-ready studio that elevates your confidence, authority, and overall viewer experience.

The power of this book lies in its adaptability. Every creator has different limitations—space, lighting conditions, background noise, budget, and the type of content they film. Instead of giving you rigid instructions, this book gives you principles, formulas, and practical layouts you can apply anywhere. You will learn how to control the direction, softness, and color of your light in ways that make your face look more flattering, your shots more cinematic, and your videos more scroll-stopping and attention-grabbing. You'll learn how to avoid the most common lighting mistakes—harsh shadows, blown-out highlights, uneven skin tone, mixed light temperatures, distracting reflections, and flat images that make you look tired or washed out. You'll also discover the simple positioning rules that professional filmmakers use to make on-camera subjects look their absolute best with minimal gear. These techniques allow you to recreate the same high-quality lighting in any room, at any time of day, regardless of your budget.

Audio is equally powerful and often more important. Many creators underestimate how deeply audio quality affects perceived professionalism and authority. A viewer may forgive imperfect visuals, but they will immediately leave if they cannot clearly hear your message. In this introduction, you'll learn why audio quality has such a strong psychological impact and why clean sound dramatically increases watch time, retention, trust, and conversions. You will understand the difference between condenser and dynamic microphones and when each one is ideal for creators filming at home. You'll learn the strategic placement methods that eliminate echo, reduce room noise, and create clean audio without needing expensive foam panels. You'll also discover how to use household items—blankets, pillows, carpets, curtains, and even clothing—to create a quiet recording environment that rivals professional studios. Clean audio is not just a technical achievement—it is an emotional bridge between you and your audience, and this book shows you how easy it is to build that bridge.

In addition to lighting and audio, you'll learn how to optimize your entire environment. This includes creating a clean, visually appealing background that supports your brand without distracting from your message. You'll explore minimalist setups, creative home office designs, aesthetic studio shelves, color-coordinated backgrounds, blurred backgrounds, and even green screen setups that allow you to create endless scenes. You will learn how to balance depth, spacing, foreground and background elements, and color temperature to make your background part of your storytelling instead of an afterthought. You'll also learn how to reduce clutter, frame your shot using clear rules, and make your space feel more polished with nothing more than a lamp, plant, or practical light source positioned in the right location. Simple design choices can make your videos look dramatically more professional with almost no cost.

One of the most important ideas in this book is the concept of strategic upgrading. Many creators buy gear randomly, following trends, YouTube reviews, or viral suggestions—not realizing that most equipment does not actually fix their biggest problems. This introduction sets the tone for a smarter, more strategic approach. You will learn the correct order to upgrade your setup so you maximize every dollar, avoid unnecessary purchases, and build a long-term system that grows with your brand. You'll learn when to invest in your first microphone, when to add your first softbox, when to switch from natural light to controlled lighting, when to treat your room for sound, and when to scale into a full studio with multi-camera setups and advanced lighting systems. This book teaches you to build your studio like a professional: intentionally, in stages, and with clarity about what matters most at each level of your growth.

By the time you finish this book, you will not only understand how to design a full studio setup on any budget—you will feel empowered to create the exact environment that elevates your confidence, improves your camera presence, and makes your content look and sound like it belongs in the top 1% of creators. You will know how to walk into any room and instantly identify where to sit, where to aim your camera, how to position your lights, and how to capture clean sound. You will know how to adjust the mood and style of your setup for interviews, tutorials, talking-head videos, product reviews, storytelling, motivational content, educational content, and more. You will have a blueprint for transforming any space—no matter how small, cluttered, dim, or loud—into a high-performing content studio that supports your brand and your message. Most importantly, you will feel in control of your environment instead of being controlled by it. Lighting and audio are the two fastest ways to elevate your authority, and now you have the knowledge to make both work for you.

This introduction is your gateway into the world of high-quality home production. Every chapter that follows will give you the practical formulas, diagrams, instructions, budget guides, store lists, and step-by-step workflows needed to create your perfect setup. Whether you begin with sunlight and a smartphone or with softboxes and an advanced microphone system, this book meets you where you are and brings you to a professional level quickly, efficiently, and without stress. Your studio is not just a physical space—it is part of your brand identity, your storytelling framework, your confidence booster, and your stage. Now it's time to prepare that stage so you can present your message to the world at the quality it deserves. Let's begin.

*Chapter 1*

# Mastering Natural Light Before Anything Else

Natural light is the most powerful, most flattering, and most accessible lighting source available to any creator, and learning to master it is the first step toward building professional-quality video production—regardless of your budget. Before you buy a softbox, LED panel, ring light, or any other lighting tool, you must understand how light behaves, how it shapes your face, and how to position yourself relative to your environment. Natural light teaches you the core principles of direction, softness, falloff, shadows, and color temperature. Every advanced lighting technique you'll learn later is simply a controlled version of what natural light already provides. In this chapter, you will learn how to use windows, open doors, skylights, and outdoor spaces to create a studio-quality look without spending a single dollar. You will also learn how the time of day affects the softness and color of the light, and how weather conditions such as clouds, rain, or bright sunlight help shape different moods for your videos. Before you invest in artificial lighting, mastering the sun is the smartest, simplest, and most cost-effective path.

Natural light works because it is a large, soft light source that wraps around the face and reduces harsh shadows. When you position yourself correctly—usually facing the window at a 45-degree angle—you create a natural Rembrandt-style lighting that adds depth, shape, and dimension to your face. This is the same lighting technique filmmakers have used for decades to make actors look dynamic and cinematic. To master natural light, start by evaluating your home or filming location. Walk around during different times of the day and observe where the light enters the room, how bright it is, how soft it feels on your skin, and whether it creates shadows that enhance or flatten your features. Avoid positioning yourself with the window directly behind you unless you're intentionally creating a silhouette or using a reflector to bounce light back onto your face. Most creators make the mistake of sitting too close to the window or too far away. Too close and the light is too strong, blowing out highlights; too far and the light becomes weak and uneven. The sweet spot is usually two to six feet from the window, depending on the intensity.

The time of day dramatically changes the quality of natural light. Early morning light is cool, soft, and gentle, making it ideal for calm or educational videos. Midday light is harsh and direct, often creating strong highlights and deep shadows that can be difficult to control—unless you use diffusion tools like sheer curtains, white sheets, or even parchment paper to soften the light. Late afternoon light, known as golden hour, produces warm, glowing tones that can make your skin look vibrant and cinematic. Filming at the same time every day gives your videos a consistent look, especially if you're creating a series, course, or ongoing content. If your home has multiple windows, test each one. North-facing windows offer the most consistent, soft, shadow-free light throughout the day. East-facing windows provide beautiful morning light; west-facing windows offer golden evening light; south-facing windows give you strong midday brightness. Your job is to test each area, film sample clips, and choose the location that matches your desired aesthetic.

One of the most powerful tools when working with natural light is diffusion. Diffusion softens the light, reduces contrast, and eliminates harsh shadows on the face. A simple white curtain, shower liner, thin bedsheet, or frosted window film can transform bright, direct sunlight into a smooth, cinematic glow. Many top creators film with nothing more than a window and a sheer curtain because it creates beautifully soft light for free. If you want more control, you can add a reflector—another inexpensive tool that bounces light onto your darker side to create balance. You can buy a reflector for under $20 or make one using a foam board covered in aluminum foil or a white poster board. Position the reflector on the opposite side of your window to create fill light, reducing shadows and giving your face a more balanced look. When you understand diffusion and reflection, you gain full control over the quality and direction of the light without buying professional equipment.

Indoor environments often require small adjustments to make natural light look as clean and intentional as possible. Start by turning off all overhead lights, lamps, and mixed light sources. Artificial house lighting has a different color temperature that usually clashes with natural light, causing your skin tone to look unnatural or orange. You want the window to be your only light source whenever possible. Once you've isolated the light, step into your frame and look for shadows under your eyes, chin, and nose. If shadows appear too strong, move slightly away from the window or adjust your angle. If the window is too bright, use diffusion. If the light is too weak, use a reflector to increase brightness. You can also modify your background to support your natural-light setup. Adding a lamp behind you can create depth, placing objects at varying distances enhances visual interest, and choosing a clean background helps ensure the focus stays on you. Always remember that natural light changes throughout the day, so what looks perfect at 10 a.m. may look different at 2 p.m. Planning your filming schedule around your best natural-light window can create consistently beautiful results.

Outdoor lighting introduces additional variables but can produce stunning results when used intentionally. Overcast days act as nature's softbox, offering soft, even lighting with no harsh shadows. Shooting in open shade—under a tree, behind a building, or under a porch—gives you soft light that wraps around the face. Avoid filming in direct midday sunlight unless you want a dramatic or high-contrast look. The sun overhead creates unflattering shadows under the eyes and nose. Instead, position yourself with the sun behind you and use a reflector or the environment to bounce light back to your face. Filming outside can also provide more cinematic backgrounds with depth, texture, and natural movement. However, wind noise, traffic sounds, and environmental distractions may require an external microphone for clean audio. Outdoor natural light is powerful but must be controlled carefully.

Once you master natural light, you gain the ability to create a professional look in any environment, even before purchasing your first artificial lighting tool. This skill ensures that when you eventually transition to softboxes, LED panels, or ring lights, you will instinctively know how to position them, diffuse them, and blend them with your surroundings. Natural light gives you the foundation to understand how shadows behave, how the angle of light affects facial features, and how to control brightness without overexposing your camera. In fact, many full-time creators still prefer natural light for its organic, flattering look. Mastering the sun is not only a technical achievement but a creative one. With practice, you will be able to transform any window into a cinematic lighting setup that elevates your presence, authority, and professionalism. This chapter builds the foundation for everything to come—budget upgrades, studio builds, three-point lighting, and full-scale production. Natural light is your first teacher, and once you learn its lessons, you will look incredible on camera in any location.

### *The Window That Changed Everything*

When Mia first decided to start her YouTube channel, she believed she needed expensive equipment—professional lights, a studio, and a high-end camera—before she could even record her first video. She spent hours watching creator videos online, seeing glamorous setups with softboxes, LED panels, RGB lights, and lavish backgrounds that looked straight out of a television studio. She felt intimidated by it all. She worked from a small apartment, sharing walls with noisy neighbors, and her budget was tight. One afternoon, she sat on her couch scrolling through reviews of lighting kits she couldn't afford when she stumbled on a simple phrase: "The best lighting in the world is free. Sit by a window." Confused but curious, she closed her laptop and walked around her apartment, studying each window as if seeing it for the first time. One was too small, one faced a brick wall, and the sliding door to her balcony seemed too bright. But something about that sentence stayed with her.

The next morning, she woke up early and decided to experiment. The sun was just beginning to rise, casting a soft, cool glow through her living room window. She pulled a chair toward it, placed her phone on a stack of books, and hit record—just to test. When she watched the footage back, she froze. The light looked soft and gentle, shaping her face naturally. Her eyes seemed brighter, her skin smoother, and the shadows were subtle and flattering. She wasn't washed out, shiny, or overexposed. She looked... professional. Better than any video she had ever filmed inside her apartment. She filmed another test clip, then another, trying different angles—45 degrees, directly facing the window, and even with the window to the side. Each clip taught her something new about how the light behaved. For the first time, she felt hopeful.

Within a few days, Mia created a routine. She learned that 9:30 a.m. was her perfect filming window: bright, soft light with just enough contrast to add dimension to her face. She began paying attention to details she had never noticed before—how shadows moved across the room, how the brightness changed when a cloud passed overhead, and how the cool tones of morning light gave her videos a calming, inviting feel. She hung a sheer white curtain over the window to soften the light, and when she did, the difference amazed her. Her footage looked like it came from a professional studio. She laughed to herself, remembering all the expensive gear she had worried she needed. She had achieved better results with nothing but sunlight and a five-dollar curtain.

One afternoon, as she prepared to film, she noticed that one side of her face looked slightly darker than the other. Instead of panicking or buying gear, she got creative. She grabbed a white poster board from her closet—leftover from a project months ago—and propped it against a chair beside her. The board acted as a reflector, bouncing light back onto the shadowed side of her face. She hit record again and gasped when she looked at the footage. The lighting was perfectly balanced—bright but soft, dimensional but flattering. In that moment, Mia realized something powerful: she didn't need to wait for better equipment to start creating professional-quality content. She just needed to understand the light already available to her.

As the weeks passed, her confidence grew. She filmed her first YouTube video, then her second, then her tenth—all using the same window setup. Her audience began leaving comments like, "Your videos look so clean!" and "Your lighting is perfect—what do you use?" She smiled every time, knowing the truth: she used the sun. She used the same natural light any creator had access to. What mattered wasn't the price of her tools—it was her understanding of how to work with them. She felt empowered, finally realizing that she didn't need a perfect home or perfect equipment to create something valuable. She just needed the courage to begin.

Then one cloudy morning, Mia panicked. The light looked dull and flat. She worried her video would look lifeless. But instead of giving up, she remembered what she had learned. Cloudy days meant softer, even light—like a giant natural softbox. She adjusted her chair, brought her reflector closer, and filmed anyway. To her surprise, the footage looked beautiful: moody, cinematic, calm. It reminded her that natural light didn't always need to be bright; it just needed to be understood. Every filming day taught her something new, and each new lesson made her feel more like a true creator.

Months later, when Mia finally upgraded to artificial lights, she found the transition easy. She already understood how light worked—how to shape it, soften it, direct it, and use it to flatter her face. When she bought her first softbox, she positioned it instinctively. When she added a second light, she knew exactly where it should go. Everything she built—every piece of her future studio—was grounded in the lessons she learned from her little window. She often thought back to that moment when she sat on her couch feeling discouraged, believing she needed thousands of dollars to start. She wished she could go back and tell herself what she now knew: the most important studio she would ever build started with nothing more than the sun. For Mia, that window wasn't just a lighting source. It was the beginning of her creative journey, the place where she discovered her voice and her confidence. It was where she filmed her breakthroughs, her mistakes, her experiments, and her first wins. That window changed everything because it taught her that the biggest barrier between her and her dreams wasn't money. It wasn't equipment. It wasn't her apartment. It was simply the belief that she needed more than she already had. Once she stepped into the light—literally and metaphorically—her entire world began to change. And from that point forward, she never forgot the lesson that launched her career: sometimes the best studio is already right in front of you. All you have to do is pull up a chair and press record.

*Chapter 2*

# Building Your First Budget Lighting Kit

Building your first budget lighting kit is one of the smartest investments you can make as a creator because it gives you consistency, control, and flexibility that natural light alone cannot always provide. While natural light is beautiful, it changes throughout the day and depends on weather conditions, window placement, and timing. Artificial lighting eliminates these limitations by giving you the ability to film anytime—morning, midnight, cloudy day, or dim room. A budget lighting kit doesn't need to be complicated, and you don't need to spend hundreds or thousands of dollars to achieve a professional look. The truth is that even a single inexpensive light can dramatically improve your video quality if you understand how to position it, soften it, and balance it with your environment. In this chapter, you will learn the exact pieces of a beginner-friendly lighting kit, how to choose the right light for your goals, and how to set it up in your home or workspace to create clean, flattering visuals on camera.

A great starting point is a simple LED panel or softbox light. LED panels are compact, adjustable, lightweight, and generally affordable. They allow you to control brightness and sometimes color temperature, giving you flexibility to match your setup to your environment. Softboxes, on the other hand, provide a much larger, softer light source by diffusing light through a large fabric panel. This mimics the softness of natural window light, making your face appear smooth and evenly lit. If you can only afford one light, choose a softbox or a wide LED panel with a diffuser because these create softer light that flatters most people on camera. Avoid small, harsh LED lights with no diffusion, as they can cast strong shadows and make your face look flat or overly bright. Look for lights with adjustable brightness so you can adapt to different filming situations, and consider whether you prefer plug-in power or rechargeable battery options.

Once you've chosen your primary light, the next step is learning how to position it correctly. Most beginners place their lights directly in front of them, which creates flat lighting with minimal dimension. While this can work for some styles of content, it's not always the most flattering. A more professional approach is to place your light at a 45-degree angle from your face and slightly above eye level, creating soft shadows that add depth and shape to your features. This is known as "key light" placement and is the standard in filmmaking and photography. You can further soften the light by adding diffusion—either the built-in diffusion panel or an additional layer like a soft cloth or parchment paper. If the light feels too strong, step away from it or reduce the brightness. Remember that your background also affects how your lighting looks; a bright background may require stronger lighting on your face, while a darker background requires less.

If your budget allows for a second light, you can elevate your setup significantly by creating a two-light system. The first light is your key light, which illuminates your face. The second light is your "fill light," placed on the opposite side of your key light to soften shadows. When using budget equipment, your fill light doesn't need to be a professional light; it can be a small LED panel, a desk lamp, or even a white reflector that bounces light from your key light. The goal is not to eliminate shadows entirely but to balance them so your face appears dimensional without being overly dramatic. If you use a reflector as fill, position it at chest height or slightly below your face to bounce soft light upward. You don't want strong overhead or upward shadows, so always adjust until the balance feels natural. Two-light setups offer control and flexibility that mimic higher-end studio systems.

For creators with a little more space, a third light called a "back light" or "hair light" can transform your setup. This light is positioned behind you, aimed toward your hair or shoulders, creating separation between you and the background. This small detail instantly gives your video a cinematic feel and prevents your head from blending into darker backgrounds.

Back lights don't require expensive gear—a tiny LED panel, clamp light, or even a lamp placed behind your chair can work. The key is to keep the light subtle; it should highlight the edges of your silhouette without overpowering your key light. A back light adds professionalism, depth, and visual interest even with the simplest budget setup.

One of the challenges when building a budget lighting kit is avoiding mixed color temperatures. Different bulbs emit different colors—some are warm (yellow/orange), others are cool (blue/white). When these mix on camera, your skin tone can look unnatural. Always choose lights with adjustable color temperature or buy bulbs labeled with the same Kelvin value. For clean, professional-looking video, aim for a consistent temperature—usually around 5000K to mimic daylight. If your room has lamps or overhead lights that you want to keep on for ambience, use bulbs with matching temperatures or turn them off entirely to avoid color conflicts. Consistency is the foundation of professional lighting, and matching the color temperature of your lights is one of the easiest ways to create cohesive visuals.

The final step in building your budget lighting kit is testing and adjusting. Film short clips at different angles, distances, and intensities to understand how changes affect your look. Move your light closer for softer, brighter illumination, or farther for a more dramatic effect. Adjust your camera settings if possible—lowering ISO will reduce noise when your lighting is strong, while increasing ISO helps in low-light environments but can make your footage grainier. Your budget kit becomes powerful when you understand how to use it intentionally. Even with inexpensive equipment, you can create stunning results by mastering position, balance, and softness. By the end of this chapter, you will know exactly how to build, set up, and optimize a budget lighting kit capable of producing professional-quality videos every time you hit record.

*The $40 Light That Started It All*

Alex had always dreamed of becoming a content creator, but every time he tried filming a video, something felt off. His face looked shadowy, grainy, or unevenly lit. Sometimes his forehead would reflect too much light from the ceiling, while the rest of his face stayed dim. Other times his background looked brighter than he did, making him appear like an afterthought in his own videos. He blamed his phone at first, then his apartment, and then his lack of experience. Like many beginners, he believed the solution was expensive equipment—hundreds of dollars' worth of studio lights, tripods, and professional gear he couldn't afford. So he filmed inconsistently, always waiting for the "perfect conditions" that never seemed to come. One night, frustrated with another failed attempt, Alex decided to search online for budget lighting options. He scrolled past dozens of pricey kits until he saw a simple product with hundreds of five-star reviews: a small LED panel for $40. The comments were filled with creators saying things like "Best investment I ever made" and "This little light changed my whole setup." Skeptical but desperate, Alex ordered it. When it arrived two days later, it felt underwhelming—lightweight, plastic, smaller than he expected. He wasn't sure something so inexpensive could make a difference. But he set it up on a cheap tripod, turned off his overhead lights, and adjusted the brightness. To his shock, his entire face lit up evenly with a soft glow he had never been able to achieve before.

Curious, he started experimenting. At first, he placed the light directly in front of him. The result was bright but flat. His face looked washed out, and his features lacked depth. He tried again, moving the light to a 45-degree angle. Suddenly, shadows formed gently on one side of his face, giving him a more cinematic, polished look. He raised the light slightly above eye level, and the difference was astonishing—his eyes sparkled, his jawline looked sharper, and the shadows fell naturally. For the first time, he looked like the creators he admired online. He couldn't believe how much of a difference one small, inexpensive light could make.

Encouraged by his progress, Alex continued to experiment. He grabbed a white foam board from his closet—leftover from a school project—and placed it on his desk beside him. The foam board acted like a reflector, bouncing light back onto the shadowed side of his face. The overall image became balanced and professional-looking. He compared the footage with his previous videos and laughed out loud. The improvement was almost unbelievable. He had spent months thinking he needed a full studio setup, yet here he was, creating a near-professional look with a $40 light and a piece of foam.

As Alex continued filming, he began to understand lighting on a deeper level. He learned how distance affected softness, how brightness impacted camera noise, and how even small adjustments could transform a mediocre shot into a stunning one. Eventually, he saved up for a second light—not a fancy one, just another small LED panel. He placed it opposite his first light and used it as a fill light. Suddenly, his face had perfect balance, with subtle shadows that looked intentional and artistic. His videos looked cleaner, more polished, and more trustworthy. His viewers noticed too. Comments began appearing: "Your lighting looks amazing now!" and "Your videos feel so much more professional lately." Alex smiled every time. He knew they had no idea how simple his setup really was.

A few months later, Alex decided to add a back light to his system. He didn't buy a new light; he simply repurposed a cheap desk lamp. He placed it behind him, angled it toward his shoulders, and adjusted the brightness until it created a soft glow around his outline. When he reviewed the footage, he was stunned. The back light separated him from the background and added a cinematic edge he had only seen in high-budget productions. The transformation made him realize something powerful: professional quality wasn't about expensive tools—it was about understanding how to use what you have.

As Alex's channel grew, so did his confidence. He upgraded his equipment slowly, but he never forgot the $40 light that started it all. It taught him that creativity mattered more than gear, that knowledge mattered more than price tags, and that progress comes from taking action—not waiting for perfect circumstances. Whenever new creators messaged him asking what equipment they needed to start, he always gave the same answer: "Buy one good, inexpensive light and learn how to use it. That is how you begin."

To Alex, that small LED panel wasn't just a piece of gear. It was the turning point in his journey—the moment he realized he didn't need a studio, a team, or a big budget to bring his ideas to life. All he needed was light, a little curiosity, and the willingness to try. And from that moment on, he filmed with confidence, knowing that every great creator starts small, every great studio begins with one simple setup, and every big success story often starts with something as humble as a $40 light.

*Chapter 3*

# Audio Quality First: Choosing the Right Microphone on Any Budget

Audio quality is one of the most important factors that determines how long someone will watch your video and whether they perceive you as a credible, trustworthy creator. While great lighting captures attention, great audio keeps attention. Poor audio—echo, background noise, thin sound, or muffled speech—creates an immediate negative impression and causes viewers to leave within seconds. Fortunately, improving audio does not require expensive equipment or technical experience. With a basic understanding of microphone types, placement techniques, and room preparation, you can achieve studio-quality sound using affordable, beginner-friendly gear. This chapter teaches you how to choose the right microphone, avoid common mistakes, and create clean, clear audio regardless of your environment or budget.

The first step is understanding the two major categories of microphones most creators use: **condenser microphones** and **dynamic microphones**. Condenser microphones are highly sensitive and capture crisp, detailed sound. They are excellent for quiet environments but easily pick up background noise—air conditioning, echo, typing, cars, pets, and voices in nearby rooms. Dynamic microphones are the opposite: they are less sensitive and focus more intensely on what is right in front of them. This makes them ideal for noisy homes, apartments with echo, or creators who film in bedrooms, offices, or kitchens. Condenser microphones typically require a cleaner acoustic environment, while dynamic microphones thrive in more challenging spaces. Understanding this difference alone can save you from choosing the wrong mic for your setup.

Next, you must choose a microphone style that fits your filming style. **Lavalier microphones**, also known as clip-on mics, attach to your shirt and capture sound close to your mouth. They are budget-friendly, work well with smartphones, and reduce room echo because the mic is so close to the sound source. They are ideal for tutorials, talking-head shots, indoor creators, or anyone moving around while filming. **Shotgun microphones** are directional mics that capture sound from in front of them while rejecting noise from the sides and back. They can be mounted on your camera, your phone, or a small boom arm just outside the frame. Shotgun mics are perfect for creators who want cleaner audio without wearing anything on their clothing. Finally, **USB microphones** are plug-and-play microphones commonly used by podcasters, streamers, and desktop creators. They can deliver excellent sound for tutorials, educational videos, voiceovers, or screen-based content—especially when paired with good positioning and basic room treatment.

Budget does not need to limit your audio quality. You can achieve professional sound for under $30 with a basic lav mic plugged into your smartphone. You can achieve even better audio for under $100 using a budget shotgun or USB mic. When choosing, avoid focusing on brand hype and instead think about your environment. If you film in a noisy home, a dynamic USB mic may be best. If you move around on camera, a lav mic is ideal. If you film with your camera on a tripod, a shotgun mic mounted just outside the frame can provide natural, crisp sound. Many creators overspend because they buy microphones based on what other creators use rather than what fits their actual recording conditions. Your goal is not to choose the "best" microphone on the market—it is to choose the microphone that works best in your unique space and filming style.

Once you have your microphone, positioning becomes the next critical step. The number one rule of great audio is simple: **the microphone must be as close to your mouth as possible without appearing in your shot.** Even the best microphone in the world will sound bad if placed too far away. For lav mics, clip them near your collarbone and conceal the cable if possible. For shotgun mics, angle them toward your mouth from above or below and position them just outside the camera frame—usually 6 to 12 inches away. For USB microphones, place them 4 to 8 inches from your mouth and speak slightly off-axis to reduce harsh pops and breathing noises. Microphone positioning can improve your audio quality dramatically more than upgrading your microphone. Many creators with budget mics sound better than creators with expensive studio mics simply because they position their microphones correctly.

The final pillar of great audio is **room treatment**, and this is where creators often struggle because they assume treatment requires expensive foam panels. In reality, the best acoustic treatment is free: soft materials. Blankets, pillows, rugs, towels, curtains, beanbags, and even clothing help absorb sound and reduce echo. Hard surfaces—bare walls, hardwood floors, empty rooms—create harsh reflections that make your audio sound hollow. Soft surfaces, meanwhile, make your voice sound warm and clean. You can dramatically improve sound quality by filming in smaller rooms, adding soft items, closing curtains, or placing two pillows on each side of your microphone. If your room echoes, set up your mic near a couch or bed. If you film near a window, close it to block noise. Acoustic perfection is not required—only enough softness to reduce reflections. Combined with good mic placement, this will create a clean, professional sound. By the end of this chapter, you will understand how to choose the right microphone on any budget, how to position it for maximum clarity, and how to optimize your environment for clean, professional audio. Whether you film in a quiet office or a noisy apartment, whether you prefer clip-on mics or desktop mics, whether you are a beginner or a seasoned creator, this chapter gives you everything you need to sound like a pro. Audio is the foundation of your authority. When your sound is clear, your message becomes powerful, and your audience stays longer, trusts more deeply, and returns more often. Great audio is not expensive. Great audio is intentional—and now you have the tools to achieve it.

### *The Day the Echo Wouldn't Stop*

When Lila filmed her first YouTube video, she thought lighting was the hardest part. She spent hours positioning her lamp, adjusting her background, and finding the perfect angle. But when she finally pressed record, everything looked decent—her face was clear, her background clean, and her framing balanced. She felt proud. Then she played the footage. Her smile faded instantly. The audio sounded like she was speaking inside a metal box. Every word echoed. The room felt hollow. Her voice sounded thin, distant, and unprofessional. She tried turning up the volume, then lowering it, but nothing helped. The echo followed her voice like a ghost. She sighed, closed her laptop, and buried her face in her hands. "Why does this sound so bad?" she muttered.

Determined to fix it, she began searching online for ways to improve audio. Every solution she found seemed expensive—microphones that cost hundreds of dollars, audio interfaces, mixers, acoustic panels, and foam that would cover an entire wall. She lived in a small apartment with thin walls and hardwood floors. She didn't have the space or the money for any of that. Frustration washed over her. "Maybe this isn't for me," she thought. But she pushed that idea aside. She was not going to quit because of echo. She kept looking. That's when she stumbled on a simple sentence in a forum: "The closer your mic is to your mouth, the better it will sound." She blinked. Could it really be that simple?

Lila took a breath and pulled out the lavalier mic she had bought months earlier for $20 but never used. She clipped it to her shirt, plugged it into her phone, and recorded another test. This time, her voice sounded closer—warmer, richer. The echo was still there, but less noticeable. Encouraged, she grabbed two pillows from her couch and placed them on each side of her camera. She recorded again. The echo softened even more, like the room was becoming smaller, friendlier. She added a blanket behind her, then placed a towel on her desk. When she listened to the new clip, she couldn't believe it. The hollow sound was nearly gone. Her voice sounded clean, present, and surprisingly professional—all without anything expensive. The next day, she tried experimenting with different microphones. She borrowed a friend's small shotgun mic and mounted it just above the frame. She positioned it 10 inches from her mouth and recorded another test. The difference was immediate. Her voice sounded crisp, natural, and clear. She didn't need to raise her volume or speak dramatically. She sounded like herself—only better. She laughed out loud as she replayed the clip. "That's me? That actually sounds good!" For the first time, she felt like a real creator.

As weeks passed, Lila became obsessed with sound. She learned that her small bedroom, filled with clothes and curtains, sounded better than her living room. She learned that speaking slightly off to the side of her microphone reduced harsh breathing noises. She learned that closing the window before filming eliminated most outside sound. She discovered that her laptop fan was ruining her audio, so she moved her mic farther away from it. Every small improvement made her audio cleaner, warmer, more inviting. She realized something profound: audio wasn't about buying expensive gear. It was about understanding sound.

Her viewers began noticing too. Comments appeared on her videos: "Your sound is so calming," "I love how clear your voice is," and "What mic do you use?" Every time someone asked, she smiled. They didn't know she filmed with a $20 lav mic, two pillows, a blanket, and three towels. They didn't know she recorded in the quietest corner of her apartment. They didn't know how many times she battled echo before mastering her setup. All they heard was clean audio that made her voice feel close, personal, and trustworthy.

Months later, when she upgraded to a dynamic USB mic, she knew exactly how to position it and how to treat her room. She knew how to listen for reflections, how to reduce noise, and how to speak with confidence. Her videos sounded polished—not because of the new mic, but because of the skills she had built from the beginning. Whenever new creators messaged her asking how she achieved such clean audio on a budget, she always said the same thing: "Start simple. Bring the mic close. Soften the room. Understand the sound." And she meant every word.

For Lila, the day the echo wouldn't stop wasn't a failure—it was the moment that changed everything. It was the day she learned that great audio wasn't reserved for professionals or creators with expensive gear. It was available to anyone willing to experiment, observe, and learn. It was the day she discovered that her voice mattered—and that with the right tools and techniques, the world would hear it clearly.

*Chapter 4*

# Transforming Any Room Into a Recording Space

Transforming a regular room into a functional, clean, professional recording space is one of the most powerful skills a creator can develop. Most people assume they need a dedicated studio, a spare bedroom, or a perfectly designed office to create high-quality content. The truth is that almost any room—small, cluttered, noisy, or imperfect—can be turned into a production-ready environment with a handful of strategic adjustments. This chapter teaches you how to evaluate a room, optimize its lighting, reduce noise and echo, remove distractions, and organize your layout so you can film confidently and consistently, even if your home seems like the "worst place" for recording.

The first step in transforming a room into a recording space is assessing its natural strengths and weaknesses. Every room has unique characteristics that influence your lighting and audio: the size of the room, the height of the ceilings, the number of windows, the type of flooring, the furniture inside, and even how the walls are constructed. Smaller rooms generally have less echo because sound reflections have less space to bounce, while larger rooms often create hollow audio unless treated with soft items. Rooms with carpet, curtains, and furniture naturally absorb sound, while rooms with bare walls and hardwood floors amplify noise and echo. Likewise, rooms with windows facing the sun offer great lighting but may also introduce glare or street noise. Your goal is not to find a perfect room—it's to understand how to work with the room you have.

Once you know the character of your room, the next step is optimizing the layout for filming. Begin by choosing a spot that provides the cleanest, least distracting background. Avoid filming directly in front of doorways, cluttered shelves, or bright windows behind you unless you purposely want a silhouette effect. Position your camera so that your background supports your brand—simple, tidy, and intentional. A plant, a lamp, a bookshelf, a piece of art, or a clean wall all make excellent backgrounds. Many creators make the mistake of thinking the background must be filled or elaborate. In reality, simplicity often looks more professional and keeps viewer attention focused on you. Once your background is chosen, place your lighting equipment accordingly. If you use natural light, set up near a window with a 45-degree angle. If you use artificial lighting, set up your key light first, then your fill light if you have one, and your back light to separate you from the background.

After your visual setup is in place, focus on audio. This is where room transformation magic truly happens. Echo and background noise are the biggest issues creators face, and both can be dramatically reduced with soft materials. If your room echoes, add blankets, towels, pillows, rugs, clothing, or curtains. Place pillows behind your camera, behind your microphone, or around your desk to absorb reflections. If your room has a closet, open it and use the hanging clothes as natural acoustic treatment. Close windows to reduce outside noise, turn off fans or AC units if possible, and avoid filming during high-noise hours in your home. If noise is unavoidable—kids, pets, traffic—use a dynamic microphone or lavalier mic to reject ambient sound. Remember that your microphone's distance from your mouth matters more than the cost of the mic itself. Bringing the mic closer improves clarity and reduces background noise significantly.

Lighting is the next component of your transformation. Your goal is to control the light in the room so your videos remain consistent no matter the time of day or weather. If you rely on natural light, use curtains, reflectors, or diffusers to shape it. If you rely on artificial light, turn off overhead lights that create shadows or color temperature issues. Use a softbox, LED panel, or ring light as your main source, and position it to flatter your face. Make sure your background is lit appropriately—either with a subtle lamp or with practical lights that add depth and visual interest. Dim or turn off lights behind you that create glare or reflections. Your lighting setup should make you the brightest part of the frame while keeping the background visible and clean. Lighting is one of the easiest ways to transform a room from ordinary to cinematic.

Beyond lighting and audio, you must consider organization and workflow. A transformed recording space is not just visually appealing—it must also be efficient. Keep your tripod in a ready-to-use position. Store your microphone within reach. Keep cables organized. Have a designated spot where your lights remain set up, even when not in use. If you constantly tear down and rebuild your setup, you'll create frustration and inconsistency, making it harder to film regularly. Even if your room doubles as a bedroom, office, or living space, find ways to minimize setup time. Use collapsible backdrops, compact tripods, and small lights that can remain in place without dominating the room. A consistent workflow boosts your motivation and lowers the resistance to creating content.

Finally, a transformed room should support your creativity and presence. Your filming space should make you feel confident, calm, and capable. It should reflect your identity as a creator. Add personal touches that make you enjoy being in the room—plants, artwork, motivational quotes, or ambient lighting. Clear unnecessary clutter from your filming area. Adjust furniture to improve your posture and energy on camera. Make sure your filming space feels like "your spot," even if it's just a corner of a larger room. With intention and a few smart adjustments, any space—tiny apartment, shared home, noisy neighborhood—can become a polished, reliable recording studio that elevates the quality of your content and your performance on camera.

By the end of this chapter, you will know exactly how to evaluate, shape, and optimize any room into an effective recording environment. You will be able to take an ordinary space and make it look and sound like a professional studio without spending much money. This transformation is not about perfection—it's about creating a space that works for you, supports your workflow, and enhances your confidence. Your room is not a limitation; it is a canvas. With the techniques in this chapter, you will learn to create a recording space that matches your vision and empowers your creative journey.

### *The Apartment Corner That Became a Studio*

When Jacob moved into his new apartment, he was excited about everything except one thing: there was nowhere that looked good for filming. Every wall was either too plain or too cluttered, every room too echoey or too dark, and the entire space felt like the opposite of a studio. He tried filming in the living room first, but the sound echoed so badly that he sounded like he was inside a tunnel. He tried the bedroom next, but the window behind him kept blowing out the background no matter how he adjusted his camera. He even tried the kitchen, but the refrigerator hummed so loudly it overpowered his voice. After several disappointing attempts, he sat on his couch, discouraged. "Maybe this apartment just isn't meant for making videos," he told himself.

But Jacob wasn't the type to quit easily. One evening, while scrolling through creator forums on his phone, he came across a post titled: "You don't need a studio. You just need a corner." Something about that line stuck with him. He stood up, walked around his apartment, and began looking at each corner differently—not as an entire room, but as a potential frame. Eventually, he found a small corner in the living room near a wall where a plant stood beside a small shelf. It wasn't perfect. It was narrow and dim, and the wall had a tiny scuff mark. But for the first time, Jacob saw possibility instead of limitations.

He started rearranging the corner. He moved the plant slightly to the right, added a small lamp, and placed a book on the shelf for visual interest. He grabbed a soft blanket and draped it over the hardwood floor to reduce echo. The room immediately sounded warmer. He grabbed two pillows from his couch and placed them behind his camera. The echo faded even more. Then he pulled his inexpensive softbox light from the closet—the one he bought months ago and barely used—and positioned it at a 45-degree angle. The corner lit up beautifully. He turned off the overhead lights, sat in front of his camera, and pressed record.

When he played the footage, his jaw dropped. The dim, ordinary corner looked cinematic and inviting. The plant added life, the lamp added warmth, and the lighting highlighted his face perfectly. His voice sounded clearer than ever before. The echo was almost gone. He couldn't believe that this tiny, overlooked corner of his apartment had become the best recording space he had ever had. He smiled to himself and recorded again. And again. For the first time since moving in, Jacob felt like a creator again.

Over the next few days, Jacob refined his setup. He added a small rug under his chair to improve the audio even more. He hung a simple poster on the wall behind him and repositioned his lamp to create depth in the background. He placed his microphone on a boom arm and angled it just outside of the frame. He opened the curtains slightly to let in a soft touch of natural light. Every adjustment made the space feel more intentional, more polished. The small corner had become his sanctuary—a place where he could focus, create, and be himself.

As his videos improved, so did his motivation. He no longer hesitated to film because everything was already set up. His camera stayed on the tripod, his light stayed in position, and his microphone was always ready. He could sit down, press record, and create. His subscribers began commenting on the improved quality of his videos: "Your setup looks amazing now!" and "Your sound is so clear!" They didn't know that his "studio" was barely three feet wide. They didn't know he used pillows, blankets, and a cheap lamp to shape the space. They didn't know how proud he was of the transformation.

One afternoon, a friend visited and noticed the setup in the corner. "Wow, this looks like a real studio," she said. Jacob laughed. "It's just a corner," he replied. But deep down, he knew it was more than that. It was a symbol of his creativity and determination. It reminded him that limitations were often illusions. He didn't need a huge space, expensive equipment, or ideal conditions. He just needed the willingness to craft something out of nothing.

Jacob's apartment hadn't changed. But because he changed the way he looked at it, he changed everything. That little corner became the birthplace of his best videos, his growing audience, and his renewed passion for creating. Every time he sat down to film, he felt grateful for the night he read the sentence that changed everything: "You don't need a studio. You just need a corner." And with that simple idea, he proved that any room—no matter how small—could become a place where big dreams begin.

*Chapter 5*

# Camera Positioning, Angles & Framing for a Professional Look

Camera positioning, angles, and framing are some of the most overlooked yet critical elements of professional video production. Even with perfect lighting and audio, poor framing can make your video look amateur, distract the viewer, or weaken your on-camera presence. Fortunately, mastering proper camera placement doesn't require expensive equipment or filmmaking experience. It simply requires an understanding of visual psychology, composition, and how the camera interprets depth, perspective, and facial features. In this chapter, you will learn exactly how to position your camera, find the most flattering angles, and frame your shot correctly so your videos instantly look cleaner, more cinematic, and more authoritative on any platform.

The first rule of professional framing is **camera height**. The camera should almost always be level with your eyes or slightly above them. When the camera is too low, it creates an unflattering upward angle that exaggerates your chin and nostrils and makes you appear less confident or less prepared. This mistake is common among creators who put their laptop on a table or phone on a low tripod. On the opposite end, when the camera is too high, it can make you look smaller or diminish your presence on screen. Eye level creates natural balance, while positioning the camera one to three inches above eye level creates a flattering "power-neutral" angle commonly used in interviews, talking-head videos, and professional content. Eye-level framing makes the viewer feel like they are face-to-face with you, building trust and connection.

The next essential element is **camera distance**. Beginners often sit too far from the camera, creating a wide shot that feels impersonal and reveals distracting background elements. The opposite mistake—sitting too close—can feel uncomfortable and distort facial features. The ideal framing for most videos is a "medium shot," also known as a "waist-up" or "chest-up" shot. This provides enough space to show natural hand movements and body language while keeping your face clearly visible. For platforms like YouTube, interviews, and educational videos, a medium or tight-medium shot conveys confidence and professionalism. For platforms like TikTok and Instagram Reels, a tighter framing—shoulders and up—helps dominate the small screen and boosts retention. Regardless of platform, you should always fill the frame with intentionality.

Composition plays a huge role in how your video feels. One of the strongest composition tools is the **rule of thirds**, where the frame is divided into three vertical and three horizontal sections. Placing your eyes along the top horizontal third creates an immediately more cinematic and professional look. Many beginners center their face vertically, placing too much empty space above their head. This is called "headroom," and too much of it can make your video look poorly framed. Position your eyes on or slightly above the top third line, leaving just enough space above your head to breathe visually. You should also avoid sitting perfectly centered in the frame unless your background is symmetrical or you want a calm, balanced aesthetic. Slight off-center placement, especially when combined with angled backgrounds, adds visual interest and depth.

Camera angles also impact how your energy is perceived. Filming straight-on conveys confidence, clarity, and authority. Filming at a slight angle—where your body faces one direction, and your head turns toward the camera—adds warmth and approachability. Wide angles can show more environment, while tight angles increase intimacy. Horizontal (landscape) orientation works best for YouTube and long-form content, while vertical (portrait) orientation is ideal for TikTok, Reels, and Shorts. Filming in the wrong orientation for your platform results in awkward cropping or wasted screen space. Always match your orientation to your distribution channel to maximize retention and visual impact.

Another crucial element of professional framing is **depth**. Depth makes your shot look cinematic, while flat backgrounds make videos appear cheap or amateur. To create depth, move your camera farther away from the wall instead of sitting directly against it. Even two to three feet of separation can change the entire look of your video. This separation allows the camera to add natural background blur (if your lens or phone allows it) and helps any background elements—lamps, plants, shelves—appear softer and more visually appealing. When filming in a small room, depth can be created by angling the camera diagonally toward a corner instead of filming straight against a wall. This simple shift instantly adds dimension.

Lighting and framing must work together. If you position your camera correctly but neglect your lighting placement, your shot may appear flat or uneven. The camera should "see into" the direction of your key light. For example, if your key light is on your right side, turn your shoulders slightly left so the light wraps across your face. Avoid having your key light behind the camera unless you want a flat, evenly lit documentary-style look. If you use practical lights in your background—a lamp, LED strip, or decorative light—they should appear soft, warm, and not overly bright. Practical lights provide separation and depth but must complement your framing, not overpower it. Finally, camera stability and clean framing complete the professional look. Always use a tripod, phone mount, or stable surface to keep the frame steady. Never record handheld for talking-head videos unless it's intentional. Make sure your frame is level; a crooked horizon line is one of the fastest ways to make a video feel unprofessional. Double-check your background for distractions such as clutter, objects sticking out of your head, harsh shadows, or bright windows. A clean frame gives the viewer nothing to focus on except you. When you control the frame with intention, everything about your video appears more polished, regardless of your budget or gear.

By the end of this chapter, you will understand how to position your camera in a way that elevates your presence, flatters your face, and maximizes viewer engagement. You will know how to create depth, maintain clean framing, and choose the right orientation and angle for every platform. These skills will take your videos from "beginner" to "broadcast-ready," even if you film with a smartphone. Camera positioning is not about equipment—it is about intention. With the techniques in this chapter, you can make any camera look like a professional tool.

### *The Angle That Changed Her Confidence*

Sofia always felt confident in person, but the moment she turned on her camera, something changed. Her face looked different, her expressions didn't feel natural, and she constantly found herself re-recording even the simplest lines. She couldn't understand why she could speak clearly to friends and coworkers but froze or looked awkward the moment she saw herself on screen. After weeks of struggling, she thought maybe she just wasn't "meant" to be on camera. But deep down, she knew she had something to share. She just needed to figure out why the camera made her look and feel so uncomfortable.

One morning, after yet another failed attempt, Sofia sat on the edge of her bed with her phone in her hands. Her latest recording made her cringe. The camera was angled slightly upward, and the shadows on her face made her look tired or stressed. Her chin looked sharper than usual, her eyes seemed smaller, and the entire video felt wrong. She sighed and whispered, "Why do I look so different?" She began researching, hoping to find an answer. Within minutes, she stumbled upon an article that said: "The camera must be at eye level or slightly above. Anything lower is unflattering." Her heart skipped. She had always set her phone on a stack of books or a low table—never once realizing the angle was working against her.

Determined to test this new discovery, she placed her phone higher—on top of a box on her dresser, raising it just an inch above her eyes. She looked at the screen and froze. She looked... better. Softer. More confident. Her eyes seemed brighter, and her jawline looked natural. She hit record and spoke a few lines. When she replayed the video, she felt a spark of excitement. "Wait... I actually look like myself," she whispered. This small adjustment gave her a sudden rush of hope.

Encouraged, she kept experimenting. She moved her chair a bit closer, then a bit farther. She noticed how distance changed the way her face looked—too close exaggerated her features, while too far made her look distant and disengaged. She eventually found a sweet spot: chest-up framing. It felt balanced, polished, and comfortable. She added a lamp in the background, moved a plant into the corner of the frame, and angled her body slightly instead of facing the camera head-on. Slowly, the shot came together. When she recorded again, the difference was unbelievable. She looked confident, calm, and more like the version of herself she wanted the world to see.

A few days later, Sofia discovered the rule of thirds. She opened her camera app's grid and positioned her eyes along the top third line. Instantly, her framing looked cinematic. She realized she had been placing her face too low in the frame, leaving too much empty space above her head. This simple tweak made her look more grounded and less awkward. For the first time, she understood how filmmakers made their shots feel intentional. She laughed to herself, feeling like she had unlocked a secret no one had ever told her.

As the week went on, Sofia continued transforming her setup. She added depth by moving away from the wall and angling her camera toward the corner of her room. She discovered how much better her background looked with just a bit of space behind her. She played with different angles—straight on for strong messages and slightly angled when she wanted a warmer, more inviting energy. Every experiment taught her something new. Every improvement made her feel more confident. She began to enjoy being in front of the camera.

A month later, Sofia watched her latest video and couldn't believe her transformation. She looked professional, poised, and authentically herself. Her subscribers noticed too. Comments flooded in: "Your videos look amazing now!" and "What changed in your setup? You look so confident!" She smiled every time she read them. They had no idea she had simply raised her camera a few inches, found the right distance, and framed herself intentionally. They didn't know how much she had struggled before or how close she had come to giving up.

One evening, after finishing a recording she felt proud of, Sofia sat in her chair and reflected on her journey. She realized that confidence on camera wasn't about being perfect or having flawless features. It was about being framed in a way that supported her presence. The camera wasn't her enemy—it was just a tool waiting to be used correctly. And with a few simple adjustments, she had turned it into an ally.

From that day forward, she reminded every new creator she met: "It's not you. It's your angle. Fix the angle, and you'll find the confidence you've had all along." And every time she repeated those words, she remembered the moment her camera finally reflected her true self—the moment everything changed.

*Chapter 6*

# Eliminating Echo, Noise & Reverb: How to Sound Studio-Quality Anywhere

Clean audio is one of the strongest indicators of professionalism in online video, and nothing ruins a recording faster than echo, background noise, and reverb. Even with great lighting and a high-quality camera, poor audio makes videos feel amateur and difficult to watch. Viewers forgive imperfect visuals, but they will not forgive harsh, hollow, noisy, or reverberant sound. Fortunately, eliminating echo and noise does not require expensive acoustic panels or soundproof rooms. With the right strategies, you can improve your audio dramatically using simple techniques and everyday items you already have at home. This chapter teaches you how to diagnose audio problems, treat your environment effectively, and record clean, studio-quality sound in any room.

The first step is understanding what causes echo and reverb. Echo occurs when sound waves bounce off hard surfaces—bare walls, wood floors, windows, countertops, or empty rooms—with nothing to absorb them. Reverb is a buildup of these reflections that makes your voice sound distant, metallic, or hollow. The larger and emptier the room, the worse the problem becomes. Conversely, small rooms with soft materials—curtains, furniture, rugs, bedding—naturally absorb sound and reduce echo. This means that even if your room seems "bad," you can still create excellent audio by adding softness and strategic acoustic barriers. You do not need professional foam; you need soft, uneven, sound-absorbing materials placed in the right locations.

The most effective way to reduce echo is to **add absorption** to the room. This can be done using rugs, blankets, pillows, towels, curtains, or clothing. If your room has bare floors, adding a rug dramatically changes your sound. If your walls are bare, hanging a blanket or curtain—even temporarily—removes harsh reflections instantly. Placing pillows or folded blankets behind your microphone absorbs sound before it reaches the walls. Filming near a bed or couch helps tremendously because large, soft furniture absorbs a significant amount of sound. Closets full of clothing act as natural vocal booths. Many creators record their voiceovers in closets not because they're soundproof, but because the soft clothing absorbs nearly all reflections. Strategic placement matters: sound treatment should be around your microphone and behind your camera, not behind you.

Once echo is under control, the next challenge is **background noise**—fans, AC units, traffic, neighbors, pets, keyboards, or refrigerator hum. You cannot fully remove these sounds with editing alone; you must prevent them from entering your recording in the first place. The most effective tool for noise reduction is microphone choice. Dynamic microphones reject ambient noise far better than condenser microphones. Lavalier microphones also help because they sit very close to your mouth. Shotgun microphones reduce noise from the sides and rear. Beyond selecting the right microphone, you can reduce noise by turning off fans, closing windows, filming during quieter hours, and placing your microphone away from noise sources. If you record near a laptop, keep distance between the microphone and the computer fan, or position the mic to the side so it doesn't capture airflow.

Another essential technique for reducing echo and noise is **microphone distance and placement**. When the microphone is close to your mouth, your voice becomes dominant in the recording, making room noise far less noticeable. This is known as increasing the "signal-to-noise ratio." Even a cheap microphone sounds excellent if placed properly. For lavalier mics, clip the mic near your chest, about 6–8 inches from your mouth. For shotgun mics, position them just outside the frame—either above or below your face—angled toward your mouth. For dynamic and USB mics, place them 4–8 inches from your mouth and speak slightly off to the side to reduce plosives and breathing sounds. Close mic technique is the single biggest factor in achieving warm, clean audio without requiring room treatment.

To further control your acoustic environment, you can create simple and effective **DIY sound barriers**. A heavy blanket hung behind your camera absorbs reflections before they reach the microphone. Two pillows placed on each side of your mic create a mini sound booth. Draping a blanket over a tall chair behind your recording area helps remove room echo. For voiceovers, placing your microphone inside a cardboard box lined with a towel creates a shockingly effective portable vocal booth. These small adjustments cost little but dramatically elevate sound quality. Even top creators use DIY methods when traveling or recording on the go because they understand that audio quality depends more on environment and technique than expensive equipment.

Finally, when your environment and mic placement are optimized, you can enhance your audio with **post-production tools**. Editing software such as Adobe Audition, Audacity, DaVinci Resolve, or even free mobile apps offers noise reduction, EQ, and compression features. Light noise reduction can remove consistent background hum. A gentle EQ boost around 3–5 kHz adds clarity to your voice. Compression makes your volume consistent and smooth. However, these tools should be used sparingly. They cannot fix bad audio—they can only polish clean audio. The true improvement comes from recording in a treated environment with proper mic technique, not from editing afterward.

By the end of this chapter, you will know exactly how to eliminate echo, reduce noise, and create clean, studio-quality audio in any room. You will understand how to transform your acoustic environment using simple materials, how to place your microphone for maximum clarity, and how to prevent unwanted noise from entering your recordings. Whether you film in a quiet office or a loud apartment, you will be equipped to produce warm, professional sound that earns viewer trust and dramatically improves watch-time. Audio is not about perfection—it is about control. And now, you have the knowledge to control your sound anywhere.

### *The Closet That Saved Her Channel*

Maya loved making videos, but every time she listened back to her recordings, she cringed at the sound. The video looked good. Her lighting was soft, her background was clean, and her camera framing was professional. But her audio? It was awful. Echo bounced around her room like she was speaking inside a cathedral. Every "S" sounded sharp, every "P" exploded, and her voice felt thin and distant. She tried re-recording dozens of times, but nothing improved. She started doubting herself.

"Maybe I'm just not meant for this," she whispered one night, staring at her screen in frustration.

One day, after posting a video she wasn't proud of, she received a comment from a viewer: "Love your content, but the echo makes it hard to listen." It hurt, but it also lit a fire inside her. She refused to let something as simple as audio ruin her passion. That evening, she grabbed her microphone, shut her bedroom door, and started experimenting. She recorded clips in every room—bedroom, kitchen, hallway, even the bathroom. Each sounded different, but none sounded good. Frustrated, she collapsed onto her bed. That's when she noticed something she had overlooked: her closet door, slightly open, full of hanging clothes that absorbed sound. She sat up slowly. Could it work?

She stepped into the closet, closed the door, sat cross-legged on the floor, and hit record. When she played the audio back, her eyes widened. It wasn't perfect, but it was the best she had ever recorded. The echo was almost gone. Her voice sounded warm and close, not hollow or sharp. She laughed out loud, covering her mouth in disbelief. "No way... the closet?" she said, smiling for the first time that day. She recorded another clip—this time holding the microphone closer. It sounded even better. Her closet, cramped and unglamorous, had become her unexpected recording sanctuary.

The next day, she went even further. She placed a pillow behind the microphone, hung a blanket across the open space above her clothes, and set her camera on a small stack of books. She felt ridiculous—but when she hit playback, the audio sounded incredibly clean. She recorded her next video entirely in the closet, and when she uploaded it, the response was immediate. Comments appeared: "Your voice sounds amazing!" and "Your audio quality improved so much!" She felt proud—prouder than she had ever felt about any video before.

But Maya didn't want to record in her closet forever. So she began learning about how sound worked: reflections, hard surfaces, soft materials. She experimented with pillows, rugs, and blankets in her bedroom. She hung a thick blanket behind her camera. She placed two cushions on each side of her mic. She laid a towel across her desk. Slowly, her bedroom began to feel less like a hollow box and more like a recording space. When she pressed record again, she couldn't believe it—her bedroom now sounded almost as good as her closet. She danced around the room in excitement.

Over the next few weeks, Maya refined her setup. She switched to a lavalier mic for convenience, learning how mic placement changed the sound. She discovered that speaking slightly off-angle reduced harshness. She found the quietest hours of the day and used them to film. She even learned basic audio editing to polish her recordings. Her sound became so clean and warm that viewers began asking what microphone she used. She always smiled at their questions. They had no idea how much she had struggled, how many nights she had spent troubleshooting, or how her closet had saved her channel.

Months later, her videos sounded professional. But Maya never forgot the breakthrough that changed everything. Whenever she saw someone online struggling with echo, she left a comment: "Try your closet. Trust me." And every time she hit 'post,' she remembered that moment she stepped into her closet for the first time—a moment of desperation that turned into a moment of discovery. A moment that proved she didn't need expensive foam or a studio. She just needed creativity, determination, and the courage to try something different.

That tiny closet—crowded, dark, and filled with clothes—became the foundation of her confidence. It reminded her that solutions often hide in unexpected places, waiting for creators willing to experiment. And as her channel grew, Maya carried that lesson with her: great sound doesn't come from perfect conditions. It comes from refusing to give up.

*Chapter 7*

# Mastering Backgrounds, Depth & Visual Aesthetics

Your background is more than the space behind you—it is part of your brand, your storytelling, and the emotional tone of your videos. A well-designed background instantly elevates the quality of your content, while a cluttered or poorly lit one distracts viewers, weakens credibility, and makes even great lighting and audio feel less professional. Fortunately, you don't need an expensive studio or large room to create visually appealing backgrounds. With intentional framing, thoughtful object placement, and an understanding of depth, you can turn any corner of your home into a clean, cinematic, and aesthetically pleasing environment. This chapter will teach you how to craft backgrounds that look professional on any budget and help your videos stand out on every platform.

The foundation of an effective background is **simplicity**. Many beginners overcrowd their frame with decorations, lights, or random objects, believing that more equals better. In reality, a professional background looks curated, intentional, and clean. You want just enough visual interest to keep the frame from feeling empty but not so much that it competes with you as the main subject. A simple plant, a lamp, a candle, a piece of artwork, or a small shelf can add the perfect amount of texture and personality. The key is to choose elements that reinforce your brand's message. For example, tech creators may choose LED accents or minimalist décor, while creators in wellness or personal development may opt for natural materials, soft colors, and calm aesthetics. Every object in your frame should have a purpose.

Depth is the secret sauce that transforms a flat-looking background into a cinematic one. When your background sits too close behind you—such as a wall directly touching your chair—the shot looks compressed and lifeless. By placing yourself two to four feet in front of the wall, you introduce depth, allowing your camera to separate you from your background. This creates natural blur, enhances your lighting, and gives a three-dimensional feel similar to high-end productions. If your space is small, you can still create depth by angling the camera toward a corner rather than filming straight against a wall. Corners naturally add dimension and make the room feel larger on camera. Even a narrow space can look expansive with the right angle.

Lighting your background correctly is another crucial step. Your background should complement your face, not compete with it. This means the background should be darker or less bright than your key light. Practical lights—lamps, LED strips, fairy lights, candles, or neon signs—can add warmth, depth, and personality. Just be careful not to create hotspots or overly bright points that distract from you. Soft, warm lighting in the background creates a welcoming atmosphere, while cooler lighting creates a sleek, modern look. Match your background lighting to your brand identity and the mood of your content. If you have a plant or shelf in your background, positioning a small light to highlight it can add a subtle but beautiful cinematic touch.

Another essential part of background design is removing distractions. Anything that draws attention away from your face reduces viewer engagement. Items like open doors, cluttered shelves, loose cables, laundry, or bright windows can ruin an otherwise great shot. A clean background communicates professionalism and attention to detail. If you cannot physically clean the entire room, clean the camera frame only. Many creators have a studio-quality background in a room that is messy everywhere except for the framed section. What matters is what the viewer sees. This allows you to create high-quality visuals even if you have a small room, shared space, or limited ability to rearrange furniture.

Color also affects the emotional impact of your videos. You don't need to repaint your walls, but you should understand how colors influence mood. Warm tones—beige, brown, gold—create comfort and approachability. Cool tones—gray, blue, white—create professionalism and clarity. Green adds freshness, purple adds creativity, and neutral tones help your face stand out. If your background is too chaotic in color, simplify it by removing items or adding neutral elements. Wearing clothing that contrasts slightly with your background helps viewers focus on you. Avoid blending into your background or wearing overly bright colors that reflect light or distract the viewer.

Finally, your background should feel like part of your brand identity. It should reflect what your content represents. If you teach productivity, your background should look clean, organized, and intentional. If you create lifestyle content, it should feel cozy and inviting. If you teach tech or digital skills, it should look modern and sharp. Your background tells a story before you even speak a word. Viewers instantly form impressions based on what they see behind you. By designing your background with purpose, you create a strong visual identity that sets you apart and builds trust from the moment your video begins.

By the end of this chapter, you will understand how to create backgrounds that look polished, cinematic, and professional without buying expensive décor or renovating your home. You will know how to use depth, lighting, composition, and clean design to shape the emotional tone of your videos. Most importantly, you will learn how to make any small space look like a professional studio using simple, budget-friendly techniques. Backgrounds are your visual signature—and now, you'll be able to create one that elevates your brand and captivates your audience.

*The Shelf That Made Him Look Like a Pro*

Caleb had been making videos for months, but no matter how hard he tried, something about his setup always looked off. His lighting was improving, his sound was much cleaner, and his camera framing was finally starting to feel right—but the background? It looked dull, flat, and lifeless. He filmed against a blank beige wall, and every time he watched his recordings, he felt uninspired. "It just looks boring," he muttered one afternoon after reviewing his latest video. He felt frustrated. He wanted his videos to feel inviting and professional, not like he was filming inside a bland, empty room.

One day, while scrolling through a creator's YouTube setup tour, Caleb paused when he noticed something interesting: the background wasn't elaborate at all. It was simple—a plant, a wooden shelf, and a warm lamp. Yet the shot looked cinematic. Caleb leaned closer to the screen, studying the setup. "Why does that look so good?" he wondered. That's when he realized the creator had placed himself several feet in front of the wall, creating depth. The shelf wasn't pressed against the wall—it was angled slightly. The lighting was soft, warm, and perfectly balanced. It clicked. His problem wasn't the beige wall; it was how he was filming against it.

Excited, Caleb stood up and grabbed his tripod. He moved it farther from the wall, then pulled his chair a few feet forward. He looked at the camera preview and felt a burst of optimism. It already looked better. He grabbed a small shelf from another room—a shelf he hadn't used in months—and placed it in the background. He added a plant from his kitchen counter and a lamp he usually kept by his bed. For the first time, his background had texture and personality. He turned on his softbox light and positioned it at an angle. When he saw the new shot, he grinned. The room looked transformed.

But Caleb wasn't done experimenting. He dimmed the background lamp slightly so it wouldn't overpower his face. He angled the plant so it created a gentle curve behind him. He moved the shelf a few inches forward to add more dimension. Every small adjustment made the shot feel more intentional. He pressed record, spoke a few lines, and then played the video back. His heart raced. This was it. This was the look he had been trying to achieve for months. Warm. Professional. Cinematic. His beige wall was still there—but now, it was part of a well-designed frame instead of a lifeless backdrop.

Over the next week, Caleb became obsessed with backgrounds. He learned how different angles created depth. He experimented with turning off overhead lights and relying on soft, practical lights behind him. He added a small LED strip behind the shelf to create a gentle glow. He rearranged books by color and height to make the background feel balanced. For the first time, he felt proud of how his setup looked. It was still simple, but it reflected his personality—calm, organized, and modern.

His viewers noticed immediately. Comments appeared on his latest video: "Your background looks awesome!" and "The new setup feels so professional!" Caleb smiled at each message. They didn't see the struggles behind the transformation—the frustration, the countless tests, or the hours spent rearranging furniture. They just saw the result: a clean, inviting background that made his videos more enjoyable to watch.

One evening, Caleb's friend visited and saw the setup. "Dude... did you remodel your apartment?" he asked. Caleb laughed and shook his head. "Nope. I just moved things around." His friend stared at the tiny corner. "This looks like a real studio," he said in amazement. Caleb felt a swell of pride. Not because the shelf or lamp were special—they were cheap, old, and simple—but because he had learned how to make them look intentional. As Caleb continued creating content, his background became part of his brand. Viewers began recognizing the plant and the warm light behind him. They found comfort in the consistency. His tiny living room corner had become a space that felt like him. Every time he pressed record, he sat a little taller, spoke a little clearer, and felt a little more confident.

Caleb eventually realized what so many creators before him had learned: you don't need a studio to look like a pro. You don't need elaborate décor or expensive furniture. You need depth, intention, and a touch of personality. That shelf—something he had nearly thrown away—became the centerpiece of his best videos. It reminded him that creativity wasn't about having more. It was about using what he already had in a way he never had before.

*Chapter 8*

# Budget Tripods, Mounts & Stabilization for Smooth, Professional Shots

Stabilization is one of the most underrated components of professional-quality video. Even with perfect lighting, audio, and background aesthetics, shaky or uneven footage immediately communicates one thing: amateur. A stable shot builds trust, elevates your production value, and allows your viewer to focus entirely on your message instead of the wobbling frame. The good news is that stability does not require expensive gimbals or high-end tripods. Even on the smallest budget, you can achieve rock-solid shots using affordable tripods, clever mounting solutions, and simple stabilization techniques. In this chapter, you'll learn exactly how to stabilize your camera, choose the right tripod for your setup, use mounts effectively, and eliminate shake forever—even if you're filming with nothing more than your smartphone.

The first step to stable video is choosing the right tripod. Tripods come in many forms: full-size tripods, tabletop tripods, phone tripods, flexible octopus-style tripods, and heavy-duty photography tripods. A full-size tripod is ideal for most creators because it allows adjustable height, stable positioning, and easy movement. When choosing a tripod on a budget, prioritize stability over features. A simple tripod with sturdy legs is better than a "fancy" tripod with weak plastic joints that wobble over time. Look for tripods with metal legs if possible, or reinforced plastic if you're on a tight budget. Make sure the tripod's height suits your filming location—many creators accidentally choose tripods that are too short, forcing them to stack books or boxes underneath. Your tripod should reach eye level or slightly above so your camera remains properly angled without additional support.

For smartphone creators, a good phone mount is essential. Many cheap mounts hold phones too loosely or grip them in a way that creates pressure on the side buttons, which can stop recordings. Look for a phone mount with a secure clamp, rubber grips, and a standard 1/4-inch screw hole so it attaches to any tripod. Some mounts include cold-shoe mounts for attaching microphones or small lights—a huge advantage for mobile creators. Flexible tripods, like octopus-style legs, are excellent for wrapping around poles, furniture, or uneven surfaces, but they should not be your primary tripod because they are less stable for long recordings. Instead, think of flexible tripods as secondary tools for creative angles or travel setups.

Stabilization also matters when shooting handheld footage. If your content includes movement—walking shots, product demos, behind-the-scenes clips, or cinematic B-roll—you can stabilize your footage affordably without a gimbal. One technique is the "tight-grip hold," where you hold your camera with both hands, keep elbows tucked in, and move your entire upper body as a unit instead of letting your hands sway. Another method is the "body sway technique," where you shift your weight slowly from foot to foot, creating smooth movement. If you film with your phone, using a simple phone grip handle can reduce shake dramatically. Even holding your phone horizontally with two hands instead of one cuts jitter significantly. If you prefer more stabilization, inexpensive gimbal stabilizers for phones now cost as little as $50–$100 and provide extremely smooth footage.

Another major element of stability is **camera placement**. A tripod may be stable on its own, but if it's placed on an unstable surface—such as a soft carpet or shaky desk—it will still wobble. Always place your tripod on a solid surface, and avoid bumping into it during recording. If your setup is in a small room where movement is unavoidable, widen the tripod legs for extra stability or place a small weight (like a bag or dumbbell) on the bottom to anchor it. If using a tabletop tripod, ensure the table is sturdy and doesn't shake when you move or type. Many creators unknowingly attach their tripod to a flimsy table, then wonder why the footage shakes every time they talk or shift their body weight.

Stabilization also applies to **camera orientation**. Switching between horizontal (landscape) and vertical (portrait) filming becomes seamless when your mount supports both orientations. If your mount doesn't rotate, you'll be forced to tilt your tripod awkwardly, which ruins stability. Always choose mounts that can flip between orientations quickly, especially if you film for TikTok, Shorts, and YouTube. Stable orientation improves both compositional balance and user experience, helping you maintain a consistent look across platforms.

The final layer of stabilization is **planning for movement**. If you do product demos, transitions, or scenes where you interact with objects, mark your position on the floor so you return to the same spot consistently. If your tripod stays in the same place every time you film, use tape to mark where the legs go. This saves setup time and ensures consistent framing. If you film outside, be mindful of wind—even light breezes can shake lightweight tripods. Anchoring the tripod with extra weight or lowering the center column helps maintain stability. For advanced creators, a monopod or shoulder mount can provide semi-stable, mobile footage for run-and-gun style shooting without needing a gimbal.

By the end of this chapter, you will know exactly how to stabilize your shots using any equipment level. Whether you film with a smartphone, DSLR, action camera, or webcam, you'll understand how to choose the right tripod, how to mount your device securely, and how to control your movement to create smooth, professional footage every time. Stabilization is not about owning expensive gear—it's about understanding the physics of camera control and using simple tools to eliminate shake. With the techniques in this chapter, your videos will immediately gain a level of polish and professionalism that sets you apart from beginners, even if your entire setup cost less than dinner.

### *The Tripod That Changed Everything*

Evan never realized how shaky his videos were until a subscriber left a comment that read, "I love your content, but the camera movement makes me dizzy." He laughed at first, thinking it was just a joke, but when he rewatched his own videos, his stomach dropped. The footage wobbled every time he moved, every time he gestured, and even when he simply breathed. His camera sat on a wobbly desk, supported by a cheap plastic stand he got for free with his phone. No wonder his shots were unstable. He felt embarrassed. He wanted his content to look smooth and professional, not like he filmed on a roller coaster.

Determined to fix the problem, Evan decided to buy his first real tripod. But when he looked online, he felt overwhelmed. Tripods ranged from $10 to $400, and he couldn't tell the difference between them. Some looked flimsy, others looked complicated. After an hour of searching, he found a simple tripod with hundreds of positive reviews for $25. It wasn't fancy—it didn't have a fluid head, counterbalance system, or carbon fiber legs—but it looked sturdy. That was enough. He ordered it and hoped for the best.

Two days later, the tripod arrived. Evan pulled it out of the box and immediately felt the difference. It was heavier than he expected—in a good way. The legs clicked into place with a satisfying sound. The mount held his phone securely without slipping or pressing the buttons on the side. He set it up in front of his recording corner and adjusted the height until the camera was at eye level. When he looked through the preview, his reflection seemed more stable, more confident. He hadn't even hit record yet, and already his setup looked better.

He filmed a short test clip. This time, the frame didn't wobble when he gestured. It didn't shake when he shifted in his chair. The footage felt solid—anchored. He played the video twice just to be sure. He smiled in disbelief. "That's all it took?" he said softly. A $25 tripod had solved a problem that had haunted his channel for months.

But Evan didn't stop there. As he grew more comfortable with his tripod, he began experimenting with angles. He raised the tripod slightly above eye level to create a flattering look. He moved it farther back to create depth in the shot. He tried filming in vertical orientation for TikTok, then switched to landscape for YouTube, amazed at how easy it was with the new mount. Each experiment made his videos look more polished.

One afternoon, he decided to film a handheld segment while walking outside. In the past, these shots were a shaky disaster. This time, he tried holding his phone with both hands, elbows tight to his sides, moving his body smoothly like he had learned from a tutorial. The footage still wasn't perfect—no handheld shot ever was—but it was significantly more stable. Encouraged, he bought a $10 phone grip handle, and the difference was night and day. He felt like he had new creative freedom.

His confidence skyrocketed. With stable shots, his videos looked cleaner. His audience stayed engaged longer. New subscribers commented on how "professional" his content looked, even though nothing in his setup cost more than a dinner out. For the first time, Evan felt like he could compete with larger creators—not because he had expensive gear, but because he understood how to use simple tools well.

One evening, his sister walked into his room while he was recording. She looked at the tripod and said, "I don't get it. It's just a stand." Evan smiled. "It's not just a stand," he said. "It's the thing that made me believe I could actually do this." She rolled her eyes, but he meant every word. That $25 tripod represented stability—not just for his camera, but for his mindset. It grounded his confidence.

Months later, Evan upgraded some of his gear, but he never replaced the tripod. It became part of his journey, a reminder of the moment everything changed. Whenever a new creator messaged him asking how to improve their videos, he always shared the same advice: "Get a stable shot first. Everything else comes after."

Because he knew the truth: sometimes, the smallest piece of equipment makes the biggest difference—not because of what it does, but because of what it unlocks.

*Chapter 9*

# Color Temperature, White Balance & Matching Your Lights for Professional Consistency

Color temperature and white balance are two of the most overlooked aspects of video creation, yet they are responsible for almost everything viewers subconsciously feel about the "look" of your footage. You can have perfect lighting placement, excellent audio, and strong framing, but if your lighting colors don't match or your white balance is off, your video will immediately look amateur. These concepts are simple to understand once explained, but they make an enormous difference in how professional and cohesive your videos appear. In this chapter, you will learn what color temperature is, how to control white balance, how to avoid mismatched lighting, and how to create a consistent visual aesthetic across all your videos—on any budget.

Color temperature refers to how "warm" (yellow/orange) or "cool" (blue/white) a light appears. It is measured in Kelvins (K). Lower numbers—around 2700K to 3500K—represent warm indoor lighting, such as lamps and tungsten bulbs. Higher numbers—5000K to 6500K—represent cool daylight.

Different lights in your environment can create different color temperatures, and when these colors mix, your video takes on an unnatural or messy look. For example, filming with a daylight softbox while a yellow table lamp is glowing behind you can create a distracting combination of blue and orange. Many beginners don't realize this is why their skin tones look strange, their background appears inconsistent, or their videos don't feel "clean." Mastering color temperature means controlling which lights are on and ensuring they share the same Kelvin range.

White balance controls how your camera interprets color temperature. If your camera's white balance is wrong, your video will look too blue, too yellow, too green, or too magenta—even if your lighting is correct. Most smartphones and cameras have automatic white balance (AWB), which performs well in simple lighting setups but struggles when color temperatures are mixed. For consistent results, set your white balance manually whenever possible. If you are filming with daylight-balanced lights (around 5600K), set your camera to the same number or use the "Daylight" preset. If you are filming with warm indoor lighting (around 3200K), select the tungsten or warm light preset. Matching the camera's white balance to your lighting ensures accurate skin tones and consistent visuals.

The simplest way to achieve professional color consistency is to **choose one color temperature and commit to it for your entire shot**. If you are filming with daylight from a window, turn off all warm indoor lights so they don't contaminate your shot. If you are filming with artificial daylight lights, close your curtains to block warm sunlight. If you prefer a warm indoor aesthetic, turn off daylight-balanced LED panels and rely on lamps, warm bulbs, and soft warm lights. Mixing the two creates unflattering results unless you understand advanced color balancing. For beginners and even most advanced creators, consistency is far more important than complexity.

Once your main light source is consistent, you can add background lights (practicals) intentionally. A warm lamp behind you can add visual interest, even if your main lighting is cool, as long as it stays subtle and doesn't affect your skin tone. The rule of thumb: your face should be lit with clean, consistent lighting; your background can add accents if they complement your aesthetic. LED strips, RGB lights, and small accent lights can also create visually appealing backgrounds, but their brightness should be low enough not to influence your overall color temperature. Keep background lights soft, warm, and dim unless your brand intentionally embraces vibrant, stylized visuals.

Another major factor in color consistency is **skin tone accuracy**. Skin tones are sensitive to small shifts in color. Too much blue makes you look sick or tired. Too much orange makes you look unnatural or overly warm. An incorrect white balance also affects the mood: cool tones feel modern, clean, and crisp, while warm tones feel inviting, calm, and cozy. Neither is "better"—but consistency creates brand identity. If you film one video that looks cool and the next looks warm, your channel loses visual cohesion. Establish your preferred temperature range early on. Many creators choose daylight-balanced lighting because it flatters skin tones and works well with most décor. Others choose warm lighting to create a lifestyle, homey, or storytelling vibe. Whatever you choose, stick with it.

If you use multiple artificial lights, ensure their Kelvin values match. Many budget lights ship with different color temperatures, even if labeled otherwise. You can tell by turning off all lights except one and comparing the color spread on the wall or your face. If one appears warmer or cooler, adjust its temperature setting (if adjustable) or replace its bulb. If your lights have no adjustment controls, you can use gels—a thin sheet of colored plastic—to change their temperature. A blue gel cools warm light; an orange gel warms cool light. Gels cost only a few dollars and allow you to unify mismatched lights in seconds.

Finally, test your setup before committing to a filming session. Record a five-second clip, then check your skin tone, background balance, and overall aesthetic. Look for unusual color casts—green-ish tint from cheap LED bulbs, purple fringing from RGB lights, or blue overcast from daylight. Adjust your white balance, brightness, and light positioning until your shot feels clean and intentional. Once you achieve consistency, your videos will immediately feel 10 times more professional. Viewers may not know exactly why your videos look so good, but they will notice the subtle polish that comes from matching your lights and properly setting your white balance.

By the end of this chapter, you will understand how to control color temperature, master white balance, unify your lighting, and create visually consistent videos that feel polished and branded. Your footage will become cleaner, your skin tones more accurate, and your overall aesthetic more intentional. These improvements require no expensive equipment—just knowledge and a few smart adjustments. Color consistency is the silent foundation of professional visuals, and now, you have the skill to control it.

*The Day Her Skin Turned Blue*

Ariana had spent all morning filming what she thought was her best video yet. Her delivery was smooth, her lighting felt perfect, and her confidence was at an all-time high. She couldn't wait to edit it. But when she transferred the footage to her laptop, her excitement vanished. Her skin looked... blue. Not slightly cool, not a little pale—full-on, winter-ice, blue. She stared at the screen in horror. "What happened to my face?" The lighting in her room had seemed fine, but the recording told a different story. Her window light was bright, her lamp was warm, and her softbox was cool. Everything had mixed together into chaos. She felt frustrated and embarrassed. She had no idea where to begin fixing it.

Determined to understand the problem, Ariana dug into research. She watched tutorials, read articles, and even posted a screenshot in a creator forum. Within minutes, someone replied: "Your color temperatures are mixed. Your white balance is confused." She blinked. Color temperatures? White balance? She had heard those terms before but never paid attention to them. She had always assumed color was something fancy creators dealt with—not something that mattered for basic talking-head videos. But now, staring at her blue-tinted skin, she realized she couldn't ignore it any longer.

The next morning, she decided to experiment. She sat in front of her camera again but started turning things off. First, the lamp. Then the overhead lights. Then she closed the curtains. Suddenly, her room looked different—cleaner, simpler, more controlled. She turned on just her softbox. The light was bright and consistent. When she checked her camera preview, her skin no longer looked blue. It looked normal. She felt hopeful. But when she pressed record and viewed the footage, something was still off. This time, her skin looked too orange. "Oh great," she muttered. But she remembered the comment from the forum: "Match your white balance to your lighting."

Taking a deep breath, she explored her camera settings. She found the white balance menu. It had options she never understood before: Auto, Tungsten, Fluorescent, Daylight, Cloudy, Custom. She selected "Daylight," matching the daylight-balanced softbox. The preview instantly shifted. Her skin tone looked real—alive, natural. Her eyes sparkled. For the first time, the colors felt right. She hit record and viewed the footage again. Relief washed over her. She looked like herself again—not blue, not orange, just Ariana.

Inspired by her discovery, she took her setup further. She added a warm lamp behind her for depth but dimmed it so it wouldn't affect her face. She tested LED strip colors behind her shelf, choosing a subtle warm glow that complemented her softbox. She experimented with different Kelvin values on her adjustable LED panel until everything matched the aesthetic she wanted: bright, clean, and modern. Each adjustment made her setup feel more intentional. She began to understand how much color influences emotion. As days passed, Ariana became obsessed with color temperature. She tested her lights at night versus daytime. She tried filming with curtains open, then closed. She compared how her skin looked under warm lighting versus cool. She learned that warm lighting made her content feel cozy and personal, while cool lighting made it feel sharp and instructional. She chose to stick with daylight-balanced lighting because it matched her personality—bright, clear, and energetic. She set her white balance manually every time. Her videos began to look incredibly consistent, almost cinematic.

Her subscribers noticed immediately. Comments flooded in: "Your videos look so clean!" and "Your colors are perfect!" One viewer wrote, "I don't know what you changed, but your videos look so professional now." Ariana smiled, remembering how she had cried over her blue skin only a week earlier. She never imagined that something as simple as color temperature could transform her entire channel.

One evening, after finishing a shoot she was proud of, Ariana sat back and looked at her setup. Her lights were simple. Her camera wasn't expensive. Her room was small. But everything worked together in harmony because she finally understood the language of color. She no longer feared the camera turning her orange, blue, or washed-out. She had control now—true creative control.

Months later, whenever new creators asked her how she achieved such clean visuals, she always said the same thing: "Learn your color temperature. Master your white balance. It changes everything." And she meant every word. Because she knew that the day her skin turned blue was the day she finally learned how to make her videos look like they belonged in the top tier of content creators—not through gear, but through understanding.

*Chapter 10*

# Building a Full Home Studio Step-by-Step (Beginner, Intermediate & Advanced Setups)

Building a home studio is one of the most transformational steps you can take as a creator. A well-designed studio gives you consistency, speed, confidence, and professional quality every time you hit record. But many people mistakenly believe a home studio requires expensive gear, a large room, or extensive technical knowledge. The truth is that you can build an effective studio at any level—beginner, intermediate, or advanced—using simple, budget-friendly tools. In this chapter, you'll learn how to build your studio in stages, starting from a zero-budget setup and evolving into a full, optimized production environment. The goal is not perfection; the goal is creating a workflow that supports your creativity, efficiency, and long-term brand.

The **beginner setup** is the foundation, and it's perfect for creators working with almost no budget. At this level, your studio relies on natural light, a simple microphone, and a clean background. Your primary light source is a window, positioned at a flattering 45-degree angle. A sheer curtain diffuses the light, and a reflector or white poster board balances shadows. For audio, a basic lavalier mic or even a smartphone headset can create surprisingly clean sound when placed correctly. Your background is a simple wall, plant, lamp, or tidy shelf. A stack of books becomes your tripod if needed. This setup teaches the fundamentals: you learn how to position light, control audio, frame your shot, and eliminate distractions. Beginners often underestimate how powerful this stage can be; many viral creators started with nothing but sunlight and creativity.

As you transition to the **intermediate setup**, your goal shifts from using what you have to gaining control. This stage introduces affordable artificial lighting, better stabilization, improved audio, and a more designed background. A softbox or LED panel becomes your main light, allowing you to film consistently at any time of day. A second small light or reflector balances shadows. A dynamic or shotgun microphone improves sound quality and reduces noise. A simple full-size tripod stabilizes your shot, and your background becomes more intentional—with depth, accent lights, and brand-aligned décor. You begin to understand how each element of your setup works together: how lighting interacts with your skin tone, how microphone placement affects clarity, and how background choices build your brand identity. The intermediate studio is where creators experience their first major leap in video quality.

The **advanced setup** focuses on efficiency, speed, and precision. This level introduces multiple lighting points, room treatment, advanced microphones, and automation. A three-point lighting setup (key light, fill light, back light) creates a cinematic look. LED panels, practical background lights, and RGB accents give you full control over color and mood. A boom-mounted shotgun mic or high-quality dynamic desk microphone captures rich, professional sound with minimal post-production. Acoustic panels or DIY sound treatment absorb reflections and create a clean audio environment. Your camera sits on a stable tripod or arm with marked positions so you never have to recalibrate framing. Your background design becomes part of your signature—recognizable and cohesive. At this stage, your entire setup can be turned on and ready to film in under two minutes, allowing you to produce content rapidly and with consistent quality.

An essential part of building a studio is **workflow optimization**. Your studio should save you time, not add complications. Keep your camera on its tripod, your microphone on its stand, and your lights positioned permanently if possible. Avoid tearing down your setup after every session. The more friction you remove, the more content you'll create. If your studio doubles as a bedroom or office, choose compact lights, collapsible softboxes, or wall-mounted fixtures to save space. Cable management also matters: tangles slow you down and add visual clutter. Velcro straps, clips, and cable sleeves keep your studio clean and functional. Every improvement in workflow translates into higher consistency and less frustration.

Another key pillar of the home studio is **sound control**. In the beginner stage, you rely on natural soft materials—pillows, rugs, blankets—to absorb reflections. In the intermediate stage, you may add foam panels or DIY acoustic treatment. In the advanced stage, you strategically place acoustic panels, corner traps, or diffusers. However, remember that perfect soundproofing is expensive and unnecessary for most creators. What matters is reducing echo and controlling your recording environment. A well-placed rug can improve audio more than a wall full of cheap foam. Good microphone technique often matters more than room treatment. Your studio must be tailored to your unique environment, not to an idealized version of what a studio "should" look like.

Lighting is the final element that ties the entire studio together. Beginners use natural light. Intermediate creators rely on one or two artificial lights. Advanced creators use layered lighting: a key light that highlights the face, a fill light that softens shadows, a back light that creates separation, and practical lights that add dimension. This layered approach allows you to shape mood, maintain consistent color temperature, and produce high-quality videos regardless of time or weather. Matching color temperatures across your lights ensures professional consistency. Advanced creators often dim their lights or use color gels to control brightness and tone. A well-lit studio makes editing easier and dramatically improves viewer engagement.

By the end of this chapter, you will understand how to build a home studio step-by-step—from bare minimum to full production environment. You will have clear guidance on choosing the right tools at each level, optimizing your space, and creating a workflow that supports long-term success. This chapter empowers you to grow with intention, increase your video quality naturally over time, and avoid unnecessary purchases. No matter your budget or experience, you now have a roadmap for building a studio that works, evolves, and ultimately becomes the backbone of your creative brand.

*The Studio That Grew With Him*

When Daniel recorded his first video, he sat in front of a window with his phone stacked on three books. He had no microphone, no tripod, no lighting—just sunlight, determination, and a vision of sharing his message with the world. His voice echoed slightly, the frame was crooked, and his background was just a bare wall. But he didn't care. He pressed record anyway. When he watched the footage, he smiled. It wasn't perfect, but it was a start. He didn't realize it then, but this tiny moment would be the seed of a studio that would grow with him, one stage at a time.

At first, Daniel filmed whenever the light looked good. Sometimes the sun was too bright, sometimes too dim, and sometimes clouds ruined everything. He would record the same clip three times just to get consistent lighting. For audio, he leaned closer to his phone to sound clearer. He hung a blanket behind the camera to reduce echo. It was simple, scrappy, and imperfect—but it worked. He was learning the fundamentals without even knowing it. He had built his **beginner setup**, and every imperfect recording taught him something new.

As his passion grew, so did his curiosity. One day, he bought a $30 softbox light. When he turned it on, everything changed. No more waiting for the sun. No more racing against the weather. For the first time, he had control. He added a cheap tripod that held his phone steadily. He bought a lavalier mic that clipped onto his shirt and instantly improved his audio. His background evolved too. He added a plant to his shelf, placed a lamp behind him for warmth, and angled his chair slightly to create depth. It wasn't fancy, but it felt intentional. His videos looked cleaner, brighter, and more professional. This was his **intermediate setup**, and he felt proud every time he pressed record. But the biggest transformation came later. Daniel decided he wanted a setup he could turn on instantly—a studio that felt like stepping onto a personal stage. He upgraded his lighting to a three-point system: a key light that wrapped gently across his face, a fill light to soften shadows, and a back light that separated him from the background. He mounted his new shotgun microphone on a boom arm just outside the frame, capturing rich, crisp sound. He rearranged his room to improve acoustics, adding a rug and some acoustic panels. His camera stayed on a dedicated tripod with tape marks on the floor so he never had to realign it. His background became part of his identity—warm lights, a bookshelf, and a plant that subtly defined his aesthetic. His filming process shrank from thirty minutes of setup to less than two minutes. He had built his **advanced studio**.

One evening, after filming a video he felt incredibly proud of, Daniel sat in his chair and looked around the room. It didn't feel like the same space where he had recorded his first shaky video. It felt like a studio. A place of creativity. A place where his ideas came alive. A place he had built piece by piece, learning through trial, error, and tiny upgrades. He realized that every stage of his studio taught him something: the beginner stage taught him resilience, the intermediate stage taught him control, and the advanced stage taught him mastery.

When Daniel's subscribers began commenting on how "professional" his videos looked, he smiled. They didn't know how humble the journey had been. They didn't know about the window that overexposed half his first video or the softbox that he initially set up backward. They didn't know about the times he filmed in his closet to avoid echo or the nights he experimented with lighting until midnight. All they saw was the final result: a confident creator with a polished studio.

Months later, Daniel received a message from a new creator: "How do I build a studio like yours? It looks impossible." Daniel replied instantly: "Start with what you have. Build one piece at a time. Your studio will grow with you." Because he knew the truth. His studio wasn't built in a day—it was built in stages, shaped by persistence and passion. And it wasn't the equipment that made him a creator—it was his willingness to start.

# Epilogue

*Your Studio, Your Story, Your Future*

A home studio is more than a collection of lights, microphones, tripods, and décor—it is the physical representation of your voice, your creativity, and your future. It grows with you, reflects you, and evolves into something uniquely yours. When you first began this journey, you may have imagined a professional studio as something distant or expensive, reserved for filmmakers, influencers, or people with big budgets and perfect spaces. But now you understand the truth: the studio you build is not about the room you have, but about the intention you bring into it. A great studio is not a location; it is a mindset of resourcefulness, clarity, and mastery. And now, you possess all of that and more.

In these chapters, you learned how to control light even if all you have is sunshine from a single window. You learned how to shape your voice with microphones, blankets, and pillows long before you ever needed acoustic panels. You discovered how camera angles can change confidence, how backgrounds speak before you do, and how color temperature gives your videos emotional tone. You learned to stabilize your shots using tripods, books, or even your own body. You discovered how to build a studio step by step—from beginner improvisation to advanced precision. And in doing so, you uncovered something far more powerful than equipment knowledge: you learned how to see your space with the eyes of a creator.

A home studio forces you to look at your environment differently. A blank wall becomes potential. A lamp becomes a practical light. A curtain becomes a diffuser. A closet becomes a recording booth. You no longer see your room as just a room—you see it as a canvas for storytelling, branding, and artistic expression. You begin to notice soft shadows, warm tones, background balance, and the subtle reflections that shape the mood of your shot. You gain control over things you once ignored. And with that control comes confidence—not just in your setup, but in your ability to create consistently, quickly, and professionally.

One of the most important lessons of this book is that your studio will never be "finished." It is a living, growing part of your creative journey. There will always be new tools, new ideas, new layouts, and new ways to improve. Maybe one day you'll add another light. Maybe you'll upgrade your microphone. Maybe you'll redesign your background to match a new personal brand. Or maybe you'll change nothing at all because you've learned how powerful simplicity can be. But no matter how your studio evolves, the knowledge you've gained here ensures you can adapt with clarity and purpose. You now understand the principles behind every decision: how light behaves, how sound travels, how framing shapes perception, and how environments influence emotion.

And what makes this journey meaningful is not just the technical mastery—it's what it unlocks inside you. With a dependable setup, you no longer waste time fixing problems or battling inconsistency. Your studio becomes a place where inspiration flows more freely. You sit down, press one button, and your ideas come alive. You record more often. You speak more confidently. You show up with energy instead of frustration. You create from a place of ease. That reliability transforms your content, your workflow, and your growth as a creator. When your creative space is optimized, you become more consistent—and consistency is what separates those who dream from those who build.

But perhaps the greatest gift of a home studio is the freedom it gives you. You don't need permission from a network, a production team, or a film crew. You don't need studio rental fees or fancy equipment. You don't wait for good weather. You don't rely on luck. You own your time, your space, and your creative process. With a small corner of your home and the knowledge in this book, you now have the power to produce professional-quality video content any day, any time, in any mood. That is true creative independence. As you arrive at the end of this book, remember that your studio does not define your creativity—you define your studio. You give it energy, personality, and purpose. You turn lights and microphones into tools that amplify your message. You bring meaning into every frame, every shadow, every highlight. And as you continue creating, your studio becomes more than just a setup—it becomes your creative home. A place where your future self will look back and say, "This is where it all started. This is where I found my voice."

And so, your journey continues. Every video you create will refine your skills. Every adjustment you make will strengthen your intuition. Every mistake will teach you something new. Your studio will evolve with your dreams, expanding as your content grows. The knowledge you've gained here is not meant for a single project. It is meant to serve you for years, through every new idea, every new video, and every new version of yourself.

Now, it's time to step into your studio—your unique, intentional, perfectly imperfect space—and bring your message to the world. Your story deserves to be heard. Your ideas deserve to be seen. And now, you have everything you need to share them with clarity, confidence, and creativity.

Your studio is ready. Your future is ready. Now press record.